ON THE HUNT WITH
GRIZZLY BEARS

SANDRA MARKLE

Lerner Publications ◆ Minneapolis

THE ANIMAL WORLD IS FULL OF PREDATORS.

Predators are hunters that find, catch, and eat other animals—their prey—to survive. Every environment has its chain of hunters. The smaller, slower, less able predators become part of the prey for the bigger, faster, more cunning ones. And everywhere, just a few kinds of predators are at the top of the food chain. These are the top predators. In wilderness areas of Alaska and the northwestern parts of Canada and the United States, one of these is the grizzly bear.

3

Why are grizzly bears one of the top predators? For one thing, the adults are big enough to hunt and kill their habitat's largest prey—elk, moose, and bison. By the time a grizzly bear is an adult, both males and females can be 6.5 feet (2 m) long and stand just over 3 to 4 feet (0.9 to 1.2 m) tall on all fours.

Adult grizzly bears also have a lot of weight to help them take down prey. A male may weigh as much as 1,700 pounds (770 kg), but a bear this big is rare. Male grizzlies are usually as heavy as 900 pounds (408 kg). Though a female is likely to only weigh half as much as a male, she is still strong enough to overpower most of the other animals sharing her home range, the area she regularly roams and hunts. And when it wants to impress, a grizzly stands up on its hind feet and becomes as much as 8 feet (2.4 m) tall—a super beast!

A grizzly bear's keen senses also make it a top predator because they help it find and track down prey. A grizzly sees about as well as a human during the day and has better nighttime vision. It also has excellent hearing thanks to its ears being sound scoops on top of its head. But its keenest sense is its sense of smell. Scientists estimate a grizzly detects smells seven times better than a bloodhound and twenty-one hundred times better than a human. No wonder this bear is a champ at sniffing out prey. When prey is gathered together in a group of animals, such as a herd of caribou, a grizzly is likely to smell them as much as 6 to 10 miles (9.6 to 16 km) away.

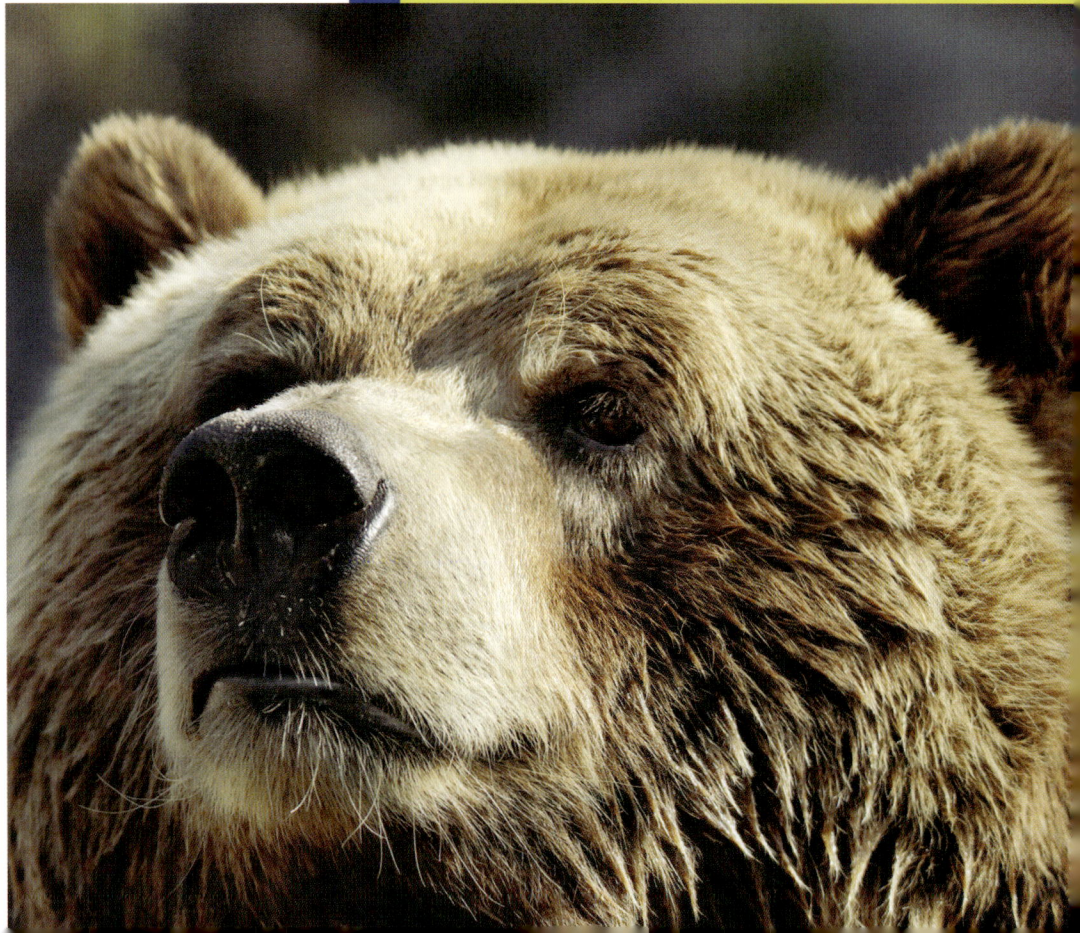

Once it senses prey, a grizzly bear can reach it quickly. Over short distances, an adult bear can run as fast as 35 miles (56 km) per hour. That's as fast as horses gallop.

But the number one reason a grizzly bear is a top predator is that it's trained by a top predator—its mother. Before her cubs were born, the female had already settled in her home range. She roamed there till she knew the area well. That way she learned to be in the right place at the right season to find berry bushes loaded with sweet, ripe fruit.

She also learned the perfect spot to catch tasty salmon in the river flowing through her home range. And she learned strategies for finding and overpowering big prey. These are all things she teaches her cubs.

The mother grizzly also makes sure her cubs get started training to hunt and become top predators during the spring and summer when prey is most plentiful. It all begins with the natural timing of when she gives birth.

A female grizzly mates sometime from mid-May to early July, but the cubs that start developing inside her stop growing after just ten days. That lets the female put all the energy from her food into building up fat reserves for winter. Early in December, once the pregnant female is in her den and dormant, her cubs continue developing. About two months later, the mother wakes up just long enough to give birth. She may have just one cub or as many as four.

At first, each cub is no bigger than an average-sized can of soup and its eyes are sealed shut. It also has no teeth, and its hair is too short to keep it warm. But it snuggles against its mother for warmth and, finding one of her nipples, nurses. This way the cubs stay warm and keep growing for the rest of the winter.

By the time the mother and her energetic, cocker spaniel–sized cubs emerge from the den, the cubs' eyes are open and they have baby teeth and fuzzy coats. But the cubs also know to stick close to their mother no matter where she goes.

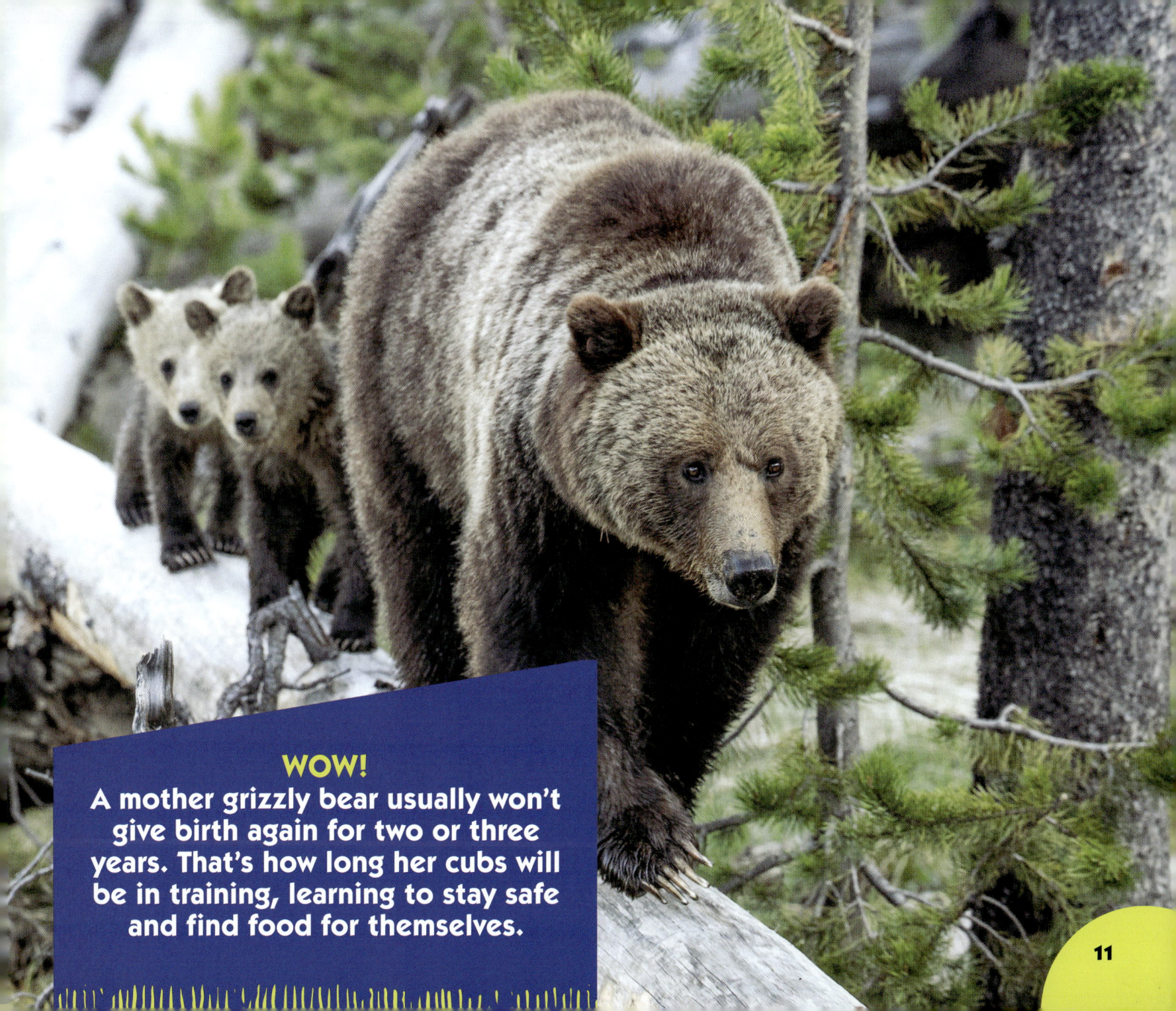

WOW!
A mother grizzly bear usually won't give birth again for two or three years. That's how long her cubs will be in training, learning to stay safe and find food for themselves.

Even after leaving the den, the cubs still mainly feed by nursing. Imagine drinking a glass full of whipping cream! A grizzly bear's milk is even thicker and creamier than that. Nursing will help the cubs grow bigger quickly. They'll keep on nursing during the two to three years they spend with their mother. But over time they will nurse less often.

Although the cubs won't get their adult teeth until they're about a year old, their baby teeth are strong enough to bite and chew. They see what their mother is eating, and like human toddlers, they nibble and try out a lot of new foods. The cubs learn where to find the food they'll eat as adults by tagging along wherever their mother goes. Watching how she catches prey is also real-life training for what they'll need to do once they're big enough to hunt on their own.

Sometimes, training to be a top predator is also play. Cubs chase one another. They play fight and jaw wrestle by grabbing a sibling's jaw and trying to pull them down to the ground. This fighting skill could help a grizzly survive when it's older. Cubs play rough, pushing and tumbling, nipping ears, and biting noses. It's a good workout that helps the young bears build strong muscles.

Keeping up with their mother while she's hunting for food helps the cubs grow stronger too. She may be leading the way, but she's always checking that her family is with her. If a cub lags behind, she huffs and grunts—sounds that have the cub running to catch up. But when one cub gets tired and squeals, its mother lets it climb on and ride until it's rested.

Besides being top predators in training, the main job for grizzly cubs is to stay alive. The cubs mimic their mother to learn how to check for danger. Like her, they stand tall on their hind legs for a better view and to catch scents in the wind. If they smell a scent they recognize as trouble, such as a male grizzly bear, the cubs know what to do.

The first rule the mother grizzly taught her cubs was to head for the treetops or the nearest cover when they need to stay safe. With trees nearby, the cubs run to the closest tree. Digging in sharp claws, they climb and don't look down until they're clinging to high branches. From this perch, the cubs watch the action as the male comes closer. Male grizzlies will kill cubs that are alone. But the mother grizzly is here!

WOW!
Scientists used X-rays of a grizzly bear's skull and computer simulations to measure an adult grizzly's bite force. They reported that it's enough to crush a bowling ball.

Roaring, the mother charges and attacks. The male (*pictured on the right*) and female (*left*) battle.

An adult's mouth is another reason a grizzly bear is a top predator. Open wide, the tips of the mother's upper and lower canines—her longest teeth—are about 5.5 inches (14 cm) apart. She bites the male's neck with tremendous force. Then she bites his shoulder. The male's jaws are powerful too, and he outweighs her. But she is fighting for her family, and he's only defending himself. When he breaks free, he hustles off.

The cubs stay put in the treetop until their mother huffs and grunts. That's their signal all is safe again. They climb down and follow her.

This time, the mother doesn't lead her cubs to find another meal or a place to nap. She takes them to the river and wades in. The cubs follow and learn a valuable lesson: soaking is a good way to escape the buzzing, biting swarms of mosquitoes. But they don't stay in the water for long. It's autumn, and food is everywhere. It's time for the grizzly bears to eat all they can before winter arrives.

WOW!

Researchers studying grizzlies in Alaska during the fall recorded them eating as much as one hundred thousand calories of food a day. That's like eating 1,282 eggs!

Day after day, the cubs wander their mother's home range with her. She eats nearly around the clock now, feasting on nuts, grubs, and lots of berries. Though the cubs still nurse, they share every meal she finds. So when their mother hunts and catches a deer, they join in by grabbing mouthfuls of this food. They just eat because it's what their mother does, but like her, they get fatter. Mother and cubs will need this stored fat to make it through the winter.

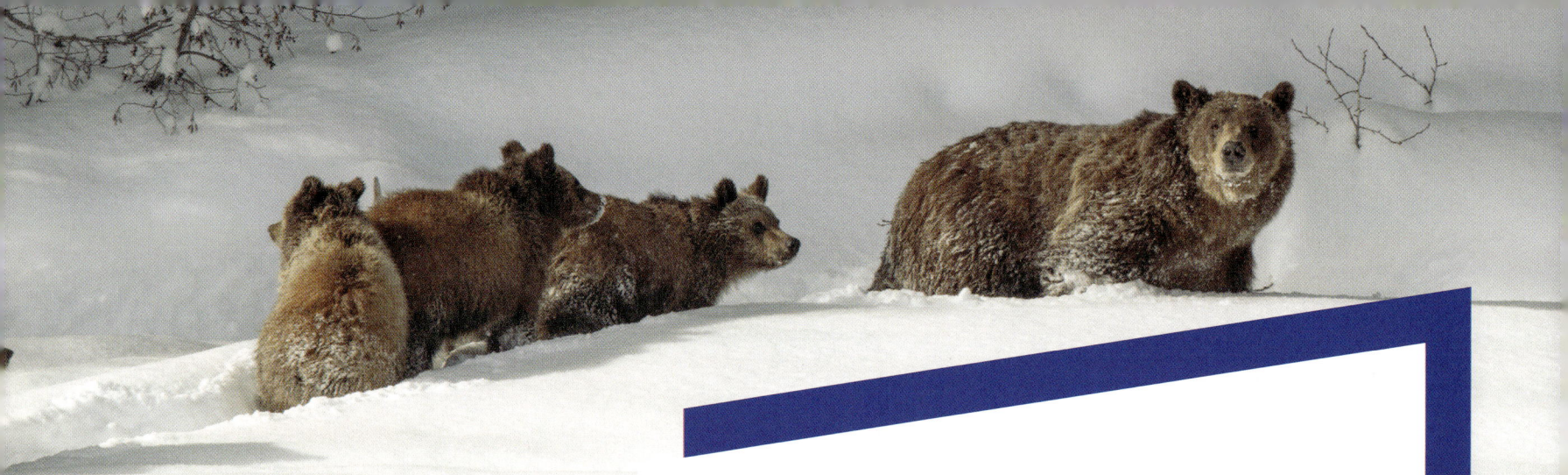

Snow falling and piling up is new to the nearly year-old cubs, but it's a familiar clue to their mother. It's time to den for winter. She leads them to a hillside sheltered by big trees. There she digs into the dirt, creating an entrance tunnel and then a rounded area where the family will curl up together. Into this chamber, the mother drags grass, tree branches, and whatever is close by that will make a nest to insulate her and the cubs from the cold ground. The entrance tunnel usually slopes uphill so the chamber is above the den's entrance. Once the opening is covered by snow, the uphill tunnel blocks cold air similar to the way storm windows help keep a house warm.

WOW!
A grizzly bear is likely to build up its wintertime nest bed to be 12 inches (30 cm) thick—or even thicker.

When she's finished den building, the mother grizzly huffs and grunts, calling the cubs to follow her inside. Curled up together, the grizzlies shift into being dormant.

The grizzly bears won't eat again until they leave the den in the spring. But once they wake up and head out, they're hungry. Over a winter of not eating, the bears have lost about 30 percent of their body weight. The nearly two-year-old cubs are ready to join their mother in hunting for food. They begin to search.

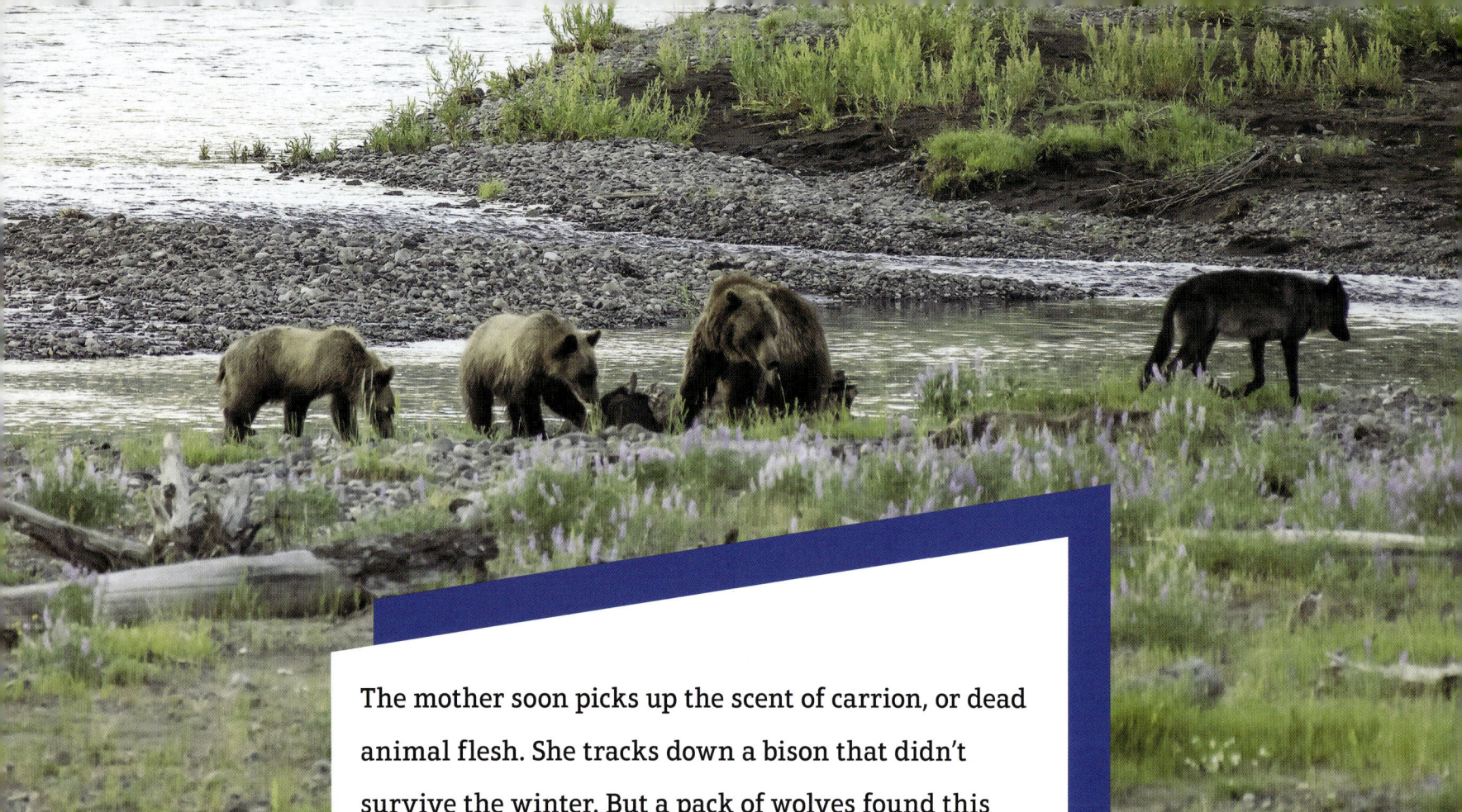

The mother soon picks up the scent of carrion, or dead animal flesh. She tracks down a bison that didn't survive the winter. But a pack of wolves found this meal first. Though at her weakest, the mother still has a top predator's gutsy attitude, and she shows her cubs how to make the most of it. With a roar, she charges in and attacks fiercely. While the wolves are still retreating, the mother huffs, calling her cubs to join her for breakfast.

With each meal, the grizzly bear family becomes stronger. The cubs are ready to move into the final stage of their top predator training. Though their mother still hunts and provides for them, they work on catching their own prey too. The cubs fail a lot at first, but every missed meal, as well as every success, is a lesson.

When one of the cubs catches a fawn without its mother's help, it's a sign. A big change is about to happen. In late June, the mother chases her cubs away. If there is more than one, they are likely to stick together for a while. Then one by one, each heads off to hunt solo, using top predator skills to thrive.

That July two males fight over the mother. She is now a lone female without her cubs. She mates with the winner and then continues roaming her home range by herself. She eats nearly nonstop through the autumn. Pregnant and heavy with stored fat, she digs a den, settles in, and becomes dormant until later that winter. One day, the female wakes just long enough to give birth to new cubs. Come spring, the mother will begin again the important job of training another generation of grizzly bear cubs to become top predators.

A NOTE FROM SANDRA MARKLE

Even top predators face risks, and global change in Earth's climate is challenging grizzly bears in two ways. First, early rises in spring temperatures act like an alarm clock for adult bears to become active and exit their dens—sometimes as much as ten days earlier than usual. For cubs just months old, that's ten fewer days to nurse and grow stronger before entering the big, challenging world.

Second, each grizzly bear's ability to thrive is affected by how climate change is altering two of its key foods—fish and whitebark pine nuts. Rising stream temperatures are killing off fish, forcing the bears to work harder to hunt other prey animals. Less frigid winters have let bark beetles survive in such numbers that the insects have destroyed many whitebark pine trees. That means fewer high-energy pine nuts for bears to pack on needed fat in the fall as they prepare for winter.

Photo by Skip Jeffery Photography

28

GRIZZLY BEAR SNAP FACTS

DIET
They mainly eat fish, deer, elk, moose, bison, and plants.

LIFE SPAN
In the wild, they live as long as thirty years.

RANGE
Grizzly bears roam wilderness areas of Alaska and the northwestern parts of Canada and the United States.

FUN FACT
One of the most recognizable parts of a grizzly bear is the big hump on its back. This large muscle powers up the bear's front legs for tearing apart rotten logs to find grubs, digging out rodents and plant roots, and digging into a hillside to create a winter den.

ADULT SIZE
Male and female grizzlies are similar lengths and can be 6.5 feet (1.9 m) long or more, but some males may weigh as much as 1,700 pounds (770 kg). Most adult males weigh about 900 pounds (408 kg). Males weigh about twice as much as females.

YOUNG
One, two, three, or even four cubs develop inside a grizzly mother's body for about 180 to 250 days before birth. They are always born during the mother's winter dormancy.

GLOSSARY

CARRION: the flesh of a dead animal

CUB: a young grizzly bear

DEN: a place where a grizzly bear finds shelter from harsh winter weather and where females give birth to cubs

DORMANT: a state in which a grizzly bear's heart rate and breathing slows so much it doesn't need to eat or pee. Being dormant allows the bear to stay inside its den throughout the winter to shelter from harsh weather conditions.

HABITAT: an animal's natural home range

HOME RANGE: the area within which a grizzly bear usually searches for food and a mate. A female raises her young in her home range.

NURSE: to feed on milk from a mother's body

PREDATOR: an animal that hunts and eats other animals

PREY: an animal that a predator catches to eat

INDEX

THE AUTHOR WOULD LIKE TO THANK DR. HARRY REYNOLDS, ALASKA DEPARTMENT OF FISH AND GAME, FAIRBANKS, ALASKA, FOR SHARING HIS ENTHUSIASM AND EXPERTISE. A SPECIAL THANK-YOU TO SKIP JEFFERY FOR HIS LOVING SUPPORT DURING THE CREATIVE PROCESS.

FOR MELISSA HAYES AND ALL THE CHILDREN AT AVERY ELEMENTARY SCHOOL IN HILLIARD, OHIO

Lerner Publications Company
An imprint of Lerner Publishing Group, Inc.
241 First Avenue North
Minneapolis, MN 55401 USA

For reading levels and more information, look up this title at www.lernerbooks.com.

Main body text set in Aptifer Slab LT Pro medium.
Typeface provided by Linotype AG.

Designer: Lindsey Owens
Lerner team: Martha Kranes

Library of Congress Cataloging-in-Publication Data

Names: Markle, Sandra, author.
Title: On the hunt with grizzly bears / Sandra Markle.
Description: Minneapolis : Lerner Publications, [2023] | Series: Ultimate predators | Audience: Ages 8–12 | Audience: Grades 4–6 | Summary: "Grizzly bears are top predators in their habitats. Their enormous size, keen senses, and ultra-sharp teeth and claws make them one of nature's greatest hunters. Learn about their lives and see them hunt down prey"— Provided by publisher.
Identifiers: LCCN 2021045331 (print) | LCCN 2021045332 (ebook) | ISBN 9781728456270 (library binding) | ISBN 9781728464404 (paperback) | ISBN 9781728462431 (ebook)
Subjects: LCSH: Grizzly bear—Juvenile literature. | Predatory animals—Juvenile literature.
Classification: LCC QL737.C27 M34525 2023 (print) | LCC QL737.C27 (ebook) | DDC 599.784—dc23/eng/20211012

LC record available at https://lccn.loc.gov/2021045331
LC ebook record available at https://lccn.loc.gov/2021045332

Manufactured in the United States of America
1-50697-50116-12/9/2021